Global Symposium on Waste Plastic 2019

Lexington, Kentucky, USA
19 – 20 September 2019

ISBN: 978-1-7138-0577-9

-Printed from e-media with permission by:

Curran Associates, Inc.
57 Morehouse Lane
Red Hook, NY 12571

AIChE
120 Wall Street, FL 23
New York, NY 10005-4020

Phone: (800) 242-4363
Fax: (203) 775-5177

www.aiche.org

Additional copies of this publication are available from:

Curran Associates, Inc.
57 Morehouse Lane
Red Hook, NY 12571 USA
Phone: 845-758-0400
Fax: 845-758-2633
Email: curran@proceedings.com
Web: www.proceedings.com

TABLE OF CONTENTS

SESSION 3: REUSE AND RECYCLE OF WASTE PLASTIC

PLENARY PRESENTATIONS

POSTER SESSION

Invited Talk: End of Life Considerations for Plastics Recyclability and Benign Degradation

- Conference: Global Symposium on Waste Plastic
- Year: 2019
- Proceeding: 2019 Global Symposium on Waste Plastic
- Group: Abstract Submissions
- Session:
  Session 1: Waste Management and Circular Economy
- Time:
  Thursday, September 19, 2019 - 10:20am-10:45am

Authors:
Urgun-Demirtas, M., Argonne National Laboratory

Currently employed standards for plastics biodegradation have major limitations that impair detailed mechanisms involved in degradation pathways. There is a need for further improvement of the current conceptual understanding of the end of life of plastics and highlighting existing knowledge gaps. Detailed knowledge of plastics degradation behavior is important not only to aid in understanding the fate of plastics but also to help develop materials and application strategies that increase plastics recovery and help inform most appropriate end-of life options to minimize environmental concerns. The main mechanisms underlying the fundamental steps in depolymerization of plastics need to be elucidated. This knowledge is crucial for biodegradable plastic producers to produce and offer consumers more responsible choice plastics and refine their processes, where end-of-life considerations such as monomer recovery or biodegradation will be an integral part of new industry practice.

Topics:
Sustainability & Environment

The Use of Supercritical Water for the Liquefaction of Polypropylene into Oil

- Conference: Global Symposium on Waste Plastic
- Year: 2019
- Proceeding: 2019 Global Symposium on Waste Plastic
- Group: Abstract Submissions
- Session:
  Session 1: Waste Management and Circular Economy
- Time:
  Thursday, September 19, 2019 - 10:45am-11:00am

Authors:
Chen, W. T., University of Massachusetts Lowell
Kai, J., Purdue University
Wang, N. H. L., Purdue University

About five billion tons of plastic waste have accumulated in landfills and the natural environment over the past 50 years. Polypropylene (PP) waste accounts for about 23% of the total plastic waste. Converting PP waste into useful products can reduce the accumulated waste and associated risks to the environment and human health. In this study, model PP was converted into oil using supercritical water at 380-500°C and 23 MPa over a reaction time of 0.5-6 hr. Up to 91 wt.% of model PP was converted into oil at 425°C with a 2-4 hr reaction time or at 450°C with a 0.5-1 hr reaction time. Higher reaction temperatures (>450°C) or longer reaction times (>4 hr) led to more gas products. The oil products consisted of olefins, paraffins, cyclics, and aromatics. About 80-90 wt.% of the oil components had the same boiling point range as naphtha (25-200°C) and heating values of 48-49 MJ/kg. Reaction pathways for converting model PP into oil under the tested conditions were proposed. Preliminary analyses indicate that this conversion process is net-energy positive and potentially has a higher energy efficiency and lower greenhouse gas emissions than incineration and mechanical recycling. The oil derived from PP has potential to be used as gasoline blendstocks or feedstocks for other chemicals.

Topics:
Sustainability & Environment

Determining Waste Plastic Generation in Developing Cities Via Geographical Analysis

- Conference: Global Symposium on Waste Plastic
- Year: 2019
- Proceeding: 2019 Global Symposium on Waste Plastic
- Group: Abstract Submissions
- Session: Session 1: Waste Management and Circular Economy
- Time: Thursday, September 19, 2019 - 11:00am-11:15am

Authors:
Joshi, C., University of Kentucky College of Engineering
Turkmani, R., University of Kentucky
Seay, J. R., University of Kentucky

With recent global attention on waste plastic generation and its accumulation on land and in oceans, several solutions for combating waste plastic accumulation have emerged. Viable solutions include ocean-cleanup efforts, reduction of waste plastic on land by recycling or incinerating, utilization of waste plastic for construction of homes, roads, and consumer goods like shoes or blankets, and complete removal of plastic from the ecosystem via pyrolysis. These solutions are altering human perception of waste plastic from garbage to a valuable resource.

However, the potential and feasibility of implementing these solutions is limited in developing countries due to lack of data on the amount of plastic waste generated in a location, such as for a city or village. Often, insufficient capital and infrastructure hinder proper collection and management of waste plastic in these regions. Thus, a tool is needed for estimating waste plastic generation in places where data is not readily available. This study performs this analysis by correlating geographical information, such as residential and commercial density, size and location of buildings to population density and income level – factors directly correlated to waste plastic generation. A case study was performed in Kampala, Uganda for development of a metric that estimates waste plastic generation by analyzing the geographical overview of the city's neighborhoods via QGIS, an open-sourced special analysis tool. Initially, input data were provided for the city's plastic generation, population density and income level for each neighborhood. In return, the metric estimates waste plastic generation for similar Sub-Saharan African cities.

Topics:
Sustainability & Environment

Integrated Framework of Plastic Footprint Toward Regional Circular Economy

- Conference: Global Symposium on Waste Plastic
- Year: 2019
- Proceeding: 2019 Global Symposium on Waste Plastic
- Group: Abstract Submissions
- Session:
  Session 1: Waste Management and Circular Economy
- Time:
  Thursday, September 19, 2019 - 11:15am-11:30am

Authors:
Fan, Y. V., Brno University of Technology
Kleme?, J. J., Brno University of Technology
Lee, C. T., Universiti Teknologi Malaysia

The plastics frequently have a short retention time (user phase) and improper disposal, not complying with the circular economy. Recycling is a promoted practice. However, the implementation is costly and requires advanced technologies due to the sensitiveness to composition (type), colour and contaminants. Some amount of plastic waste is claimed to be recycled by exporting into the developing countries. Energy recovery can be a viable solution despite some barriers, e.g. off-gas cleaning, the capacity of infrastructure, investment cost. The accounting methodology in assessing the footprint has a direct impact on the decision making. A comprehensive and consistent quantitative framework still needs more development. This study aims to highlight the critical issues in assessing Plastic Footprint and propose an integrated framework to facilitate plastic waste allocation for recovery. The Plastic Footprint introduced by Boucher et al. at the International Union for Conservation of Nature, focuses on leakage and litter, can be a basis for the integrated framework development. This work suggests the integration of plastic footprint with Energy and GHG Footprints to prevent the possibility of footprints shifting as well as to facilitate the plastic waste allocation better. The novel approach is supported by a graphical visualisation. This framework is also enabling the footprints comparison among different materials e.g. paper, glass and metal. The circular economy is not confined to a country border. The plastic waste import and export activities could be complementary when sharing resources, treatment facility, capacity and infrastructure, supply chain and the demand for recovered products are optimised.

Topics:
Environmental Sustainability

Health Risk Assessment of Plastic Waste in the Frame of the Circular Economy

- Conference: Global Symposium on Waste Plastic
- Year: 2019
- Proceeding: 2019 Global Symposium on Waste Plastic
- Group: Abstract Submissions
- Session:
  Session 1: Waste Management and Circular Economy
- Time:
  Thursday, September 19, 2019 - 11:30am-11:45am

Authors:
Sarigiannis, D., Aristotle University of Thessaloniki
Gotti, A., Aristotle University of Thessaloniki
Gypakis, A., Aristotle University of Thessaloniki
Karakitsios, S., Aristotle University of Thessaloniki

In this study we have developed an innovative tool for integrated health risk assessment of plastic waste. The INTEGRA LCA software couples the integrated external and internal exposure assessment capabilities of the INTEGRA computational platform with life cycle impact assessment. The integrated software platform allowed us to perform a first-of-its-kind analysis of adverse health outcomes attributable to chronic exposure to persistent organic pollutants associated with plastic material use and disposal.Our analysis focused on plastic waste generated in the two main metropolitan centers in Greece, Athens and Thessaloniki.

The coupled integrated exposure and life cycle assessment methodology developed in this study and translated into the INTEGRA LCA platform is a significant step towards the direction of comprehensive, precise and transparent estimation potential health risks associated with use, management and disposal of plastics in urban settings. Our first results in the two major Greek metropolitan centers indicate the need to optimize the management of plastic waste in the country and to enhance the awareness of consumers with regard to the long-term adverse health effects of chronic exposure to plastic components. The incorporation of life cycle analysis produces different conclusions than a simple environmental impact assessment based only on estimated or measured emissions. Taking into account the overall life cycle of both the waste streams and of the technological systems and facilities envisaged under the plausible scenarios analyzed herein, alters the relative attractiveness of the solutions considered and enhances the robustness of the health impact assessment.

Topics:

Safety & Health
Sustainability & Environment

The Plastic Pileup in Thailand: Illuminating Local Voices and Perspectives on the Causes and Grassroots Solutions

- Conference: Global Symposium on Waste Plastic
- Year: 2019
- Proceeding: 2019 Global Symposium on Waste Plastic
- Group: Abstract Submissions
- Session:
  Session 1: Waste Management and Circular Economy
- Time:
  Thursday, September 19, 2019 - 11:45am-12:00pm

Authors:
Meyer, O., University of Kentucky

Plastic waste is accumulating at an unprecedented rate and threatens human and ecosystem health around the world. The plastic waste issue is especially pervasive in Thailand, one of the six most significant contributors to marine plastic pollution (Jambeck et al., 2015). While international media covers Thailand's ranking and the plastic-induced deaths of megafauna on its shores, Thai regulatory institutions, grassroots environmental organizations, and residents living and working near plastic waste sites are left out of these discourses. Recognizing the importance of Thai knowledges and experiences has the potential to improve how we respond to the plastic waste issue, but not enough is known about their perspectives. As attention centers on this country, we have yet to hear from the voices of Thai people. This research involves a qualitative ethnographic study of Thai people's lived experience with and perspectives on the sources and effects of plastic waste, the power relationships underlying discourses that inform the issue, and Thai responses and agency. Through discourse analysis of ten in-depth semi-structured interviews, I engage with those involved in plastic waste governance, production, recycling, and environmental activism in the Bangkok metropolitan area. Preliminary findings suggest that those in power are reticent to alleviate the plastic pileup through measures that would challenge plastic production. These groups contest Thailand's ranking in the Jambeck et al. study while grassroots organizers call for western-derived solutions and regulatory measures. These proposed solutions must be considered in the local context.

Topics:
Sustainability & Environment

Invited Talk: Polymer Characterization of Plastic Debris from the Central Pacific

- Conference: Global Symposium on Waste Plastic
- Year: 2019
- Proceeding: 2019 Global Symposium on Waste Plastic
- Group: Abstract Submissions
- Session:
  Session 2: Ecosystem Impacts and Remediation Strategies
- Time:
  Thursday, September 19, 2019 - 1:30pm-1:55pm

Authors:
Beers, K. L., NIST

Identification of ocean debris is one focus area of an emerging Circular Economy program at the National Institute of Standards and Technology (NIST) that includes a new network of academic centers dedicated to plastics recycling technology and training. This talk will provide a brief overview of NIST's role and capabilities, and then focus on early work facilitating the development of PCA tools for FT-IR through more explicit identification of samples via high throughput, multi-detection size exclusion chromatography. Results of preliminary investigations of surface degradation and other possible marine exposure effects will also be presented.

Topics:
Sustainability & Environment

Invited Talk: Waste Valorization: an Approach to Advanced Functional Materials

GLOBAL SYMPOSIUM ON WASTE PLASTIC

- Conference: Global Symposium on Waste Plastic
- Year: 2019
- Proceeding: 2019 Global Symposium on Waste Plastic
- Group: Abstract Submissions
- Session:
  Session 2: Ecosystem Impacts and Remediation Strategies
- Time:
  Thursday, September 19, 2019 - 1:55pm-2:20pm

Authors:
Nagarajan, R., University of Massachusetts Lowell
Xia, Z., University of Massachusetts Lowell
Yu, S.
Kulkarni, S., University of Massachusetts Lowell
Mosurkal, R., CCDC SC
Kumar, J., University of Massachusetts Lowell

Advanced functional materials development is increasingly driven by new requirements as well as sustainability considerations. Waste co-products can serve as sustainable and cost-effective feedstock for the production of the next generation of safer advanced functional materials. This concept is illustrated with two classes of waste co-products that can be efficiently converted into advanced functional additives.

The development of non-halogenated flame retardant materials (FR) using co-products obtained from the food industry will be presented. It has been demonstrated that combinations of naturally occurring phenols and phytic acid can be utilized to yield additives that have excellent char forming and fire retardant properties.

Biobased and bio-inspired surfactants are the second class of materials that will be discussed. Non-ionic surfactants such as nonylphenol ethoxylates (NPE), are being phased out due to their aquatic toxicity. A new class of surfactants based on polysaccharides obtained predominantly from fruit waste has been developed recently. These surfactants derived from waste exhibit surface activity similar to that of NPE.

These two examples demonstrate the possibility of utilizing co-products to develop advanced functional materials that serve as sustainable alternatives to toxic materials such as halogenated flame-retardants and toxic surfactants. The benefits for the pursuit of materials research using waste feedstock transcends sustainability considerations and often opens exciting possibilities, making it worthwhile from a scientific and commercial perspective.

Topics:
Design/Green Engineering
Materials

Plastic Pollution in the Oceans

- Conference: Global Symposium on Waste Plastic
- Year: 2019
- Proceeding: 2019 Global Symposium on Waste Plastic
- Group: Abstract Submissions
- Session:
  Session 2: Ecosystem Impacts and Remediation Strategies
- Time:
  Thursday, September 19, 2019 - 2:20pm-2:35pm

Authors:
Peters, R. W., University of Alabama at Birmingham
Sultana, T., University of Alabama at Birmingham
Chy, A. B. M. T. U., University of Alabama at Birmingham
Nabors, M. C., University of Alabama at Birmingham
Patel, M., University of Alabama at Birmingham

During this past spring semester, we had a new course offered on "Environmental Disasters – Lessons Learned". The graduate students selected the topic of plastic pollution in the oceans for their class project.

The threat to marine life in the oceans comes in various forms such as dumping of waste, over exploitation and harvesting, pollution, invasive species, land reclamation, dredging, and global climate change. One particular type of human impact that represents major threat to marine life is the pollution by plastic debris. Plastic has become one of the most common forms of pollution in the ocean and is causing problems to become more and more severe each day, impacting both the natural environment and human life.

Plastic pollution has negative impact on our oceans and the health of wildlife, causes harm to the creatures that live in the ocean (from coral reefs to turtles gagging on straws, to whales and seabirds). These creatures starve because their bellies become jammed with plastic. Another recent concern regarding plastic pollution is the impact on food webs and marine ecosystems. The concentrated toxins of plastic materials might be delivered to animals via ingestion and transferred to their food webs. At least 26 species of cetaceans have been documented to ingest plastic debris.

This paper/presentation addresses how plastics end up in oceans, a brief history of plastic production and pollution, ecological effects of plastic pollution in the oceans, environmental impacts, health impacts, economic implications, ethical risks and concerns, and a discussion of lessons learned.

Abstract:

P582712.pdf
Topics:
Sustainability & Environment

Plastic Derived Fuel Oil (PDFO) Vs Conventional Diesel: A Qualitative and Quantitative Analysis in a Developing Country Context

- Conference: Global Symposium on Waste Plastic
- Year: 2019
- Proceeding: 2019 Global Symposium on Waste Plastic
- Group: Abstract Submissions
- Session:
  Session 2: Ecosystem Impacts and Remediation Strategies
- Time:
  Thursday, September 19, 2019 - 2:35pm-2:50pm

Authors:
Kizza, R., Makerere University
Seay, J. R., University of Kentucky
Banadda, N., Makerere University
Browning, S., University of Kentucky College of Engineering

Due to depleting fossil fuel sources, Plastic Derived Fuel Oil (PDFO) has gained so much attention over the past decade as an alternative to Conventional fossil diesel. Although several researchers have reported the properties of PDFO and recommended its usage in diesel engines, no study has done a detailed qualitative and quantitative comparison with conventional diesel. This study aims to qualitatively and quantitatively assess PDFO produced from pyrolysis of waste plastic based on standard recommended tests for a diesel in a developing country. Waste polyethylene (HDPE and LDPE), Polypropylene (PP) and Polystyrene (PS) were pyrolyzed independently in a 3T UKATS processor designed and fabricated for developing countries to produce PDFO. The produced PDFO and conventional diesel (bought from a gas station) were analyzed for distillation test, density, flash point, marker concentration, copper strip corrosion, energy content, and viscosity, and then compared with the standards. The energy recovery potential of PDFO was computed based on the Non-Recyclable Plastic (NRP) collected and dumped at the landfill whereas for conventional diesel, it was based on the production data from the Ministry of Energy and Mineral Development. Results of the qualitative analysis indicate that PDFO is a potential substitute for conventional diesel and can be used as a stand-alone fuel or as a blend. The energy recovery calculations reveal that Uganda produces waste plastic that can be converted into energy enough to serve the entire population of Kampala city.

Topics:
Fuels

Permanent Disposal of Plastic Waste By Carbon Sequestration

- Conference: Global Symposium on Waste Plastic
- Year: 2019
- Proceeding: 2019 Global Symposium on Waste Plastic
- Group: Abstract Submissions
- Session:
  Session 2: Ecosystem Impacts and Remediation Strategies
- Time:
  Thursday, September 19, 2019 - 2:50pm-3:05pm

Authors:
MacKay, I., Yankee Scientific, Inc.
Guyer, E., Yankee Scientific, Inc.

Carbon capture and sequestration is a technically feasible and economically attractive answer to the unsolved and growing problem of the responsible management of low-grade, post-consumer plastic waste in a world increasingly focused on the environmental impacts of carbon in the atmosphere. While reduced use and recycling of plastic can contribute to limiting the impact of plastic production and use on the environment, new solutions to the problem of accumulating plastic waste are needed. Solutions for plastic waste disposal must be scalable to the magnitude of the problem and will be acceptable, in the future, only if they do not contribute carbon emissions to the atmosphere. This paper presents the technology and economic rationale for deployment of processing plants that convert plastic waste to carbon dioxide for permanent sequestration in deep geologic structures. Carbon capture and storage technology has been actively pursued in connection with power generation around the world but has met economic challenges in that application. The paper discusses how and why these developments in carbon capture technology present a sustainable, affordable, and conclusive answer to the problem of plastic waste management.

Topics:
Carbon Capture & Storage
Process Design & Development

Electrospinnig of Recycled High Impact Polystyrene Nanofibers for Oil Spill Removal

GLOBAL SYMPOSIUM ON WASTE PLASTIC

- Conference: Global Symposium on Waste Plastic
- Year: 2019
- Proceeding: 2019 Global Symposium on Waste Plastic
- Group: Abstract Submissions
- Session:
  Session 2: Ecosystem Impacts and Remediation Strategies
- Time:
  Thursday, September 19, 2019 - 3:05pm-3:20pm

Authors:
Ahmed, A. M. H., Bureau Veritas SA

The conventional techniques used for oil spill decontamination have many disadvantages as high operating cost, generation of secondary pollutant and complex operation techniques. Clean up and recovery processes from an oil spill are difficult and depend upon many factors, including the type and the viscosity of the oil spilled, the temperature and the type of water. Electrospun nanofibers have started to show great attention in the last three decades, they are characterized by high surface area to volume ratio and better mechanical properties. Oil sorbent nanofibers were prepared from recycled high impact polystyrene (HIPS) by the electrospinning process. The fibers surface morphology, crystallinity, hydrophobicity, and the sorption capacity over potable water and sea water were studied. The results revealed that the contact angle of HIPS fibrous sorbent reached 153 degrees and the sorption capacity for crude oil, gas oil and motor oil were 50, 62, 125 g/g, respectively. The sorption capacity is higher than regular general purpose polystyrene (PS) nanofibers making it a highly promising material for oily water purification.

Topics:
Nanomaterials

Recycling of PET through the Manufacturing of Carbon Nanofibers

- Conference: Global Symposium on Waste Plastic
- Year: 2019
- Proceeding: 2019 Global Symposium on Waste Plastic
- Group: Abstract Submissions
- Session:
  Session 2: Ecosystem Impacts and Remediation Strategies
- Time:
  Thursday, September 19, 2019 - 3:20pm-3:35pm

Authors:
Svinterikos, E., United Arab Emirates University
Zuburtikudis, I., Abu Dhabi University
Al Marzouqi, M., United Arab Emirates University

Poly(ethylene terephthalate) is one of the most widely used commodity plastics and the main material for the manufacture of water bottles. Hence, the development of alternative uses of waste PET will have a positive impact on growing environmental concerns.

Our strategy consists of combining recycled PET with lignin for the manufacture of carbon nanofibers. Lignin and PET are miscible and can be blended by dissolving them in the same solvent, as it is demonstrated in our published work. Using the electrospinning technique, the blend is spun into fibrous mats of controllable average fiber diameters in the submicron- or nano-scale. The presence of PET is crucial for the formation of fibers, as lignin alone cannot be electrospun successfully. Finally, the electrospun lignin/PET nanofibers are thermally treated under inert atmosphere at elevated temperatures (>800 ^{o}C) for transforming them into carbon nanofibers.

Various important features of the carbon fibrous mats, such as the morphology and the porosity, are determined by the average fiber diameter of the precursor electrospun mats, as well as by the relative percentage of PET in the blend. Carbon nanofibers of minimum average diameter (close to 100 nm) can be prepared successfully only when the PET percentage is as low as 10 % w/w. When the percentage of PET is much higher, then it is not feasible to produce carbon nanofibers. However, a higher percentage of PET (around 50 %w/w) leads to much higher microporosity, as the PET disintegrates during carbonization leaving empty molecular space inside the carbon nanofibers.

Chemical Recycling of Waste PET with Super-Acid Catalysts

- Conference: Global Symposium on Waste Plastic
- Year: 2019
- Proceeding: 2019 Global Symposium on Waste Plastic
- Group: Abstract Submissions
- Session:
  Session 3: Reuse and Recycle of Waste Plastic
- Time:
  Friday, September 20, 2019 - 11:10am-11:25am

Authors:
Coleman, M. R., University of Toledo
Abedsoltan, H., University of Toledo
Alsenani, H., University of Toledo

This study focused on developing and applying novel super-acid catalysts to chemical recycling of waste polyethylene terephthalate (PET). Acid hydrolysis reaction of PET produces terephthalic acid (TPA) and ethylene glycol (EG). The effect of 8M H_2SO_4 as acid catalyst at 150 ^oC was measured to establish a baseline reaction kinetics. The goal of this work is to develop recoverable catalysts with increased activity in hydrolysis of PET to avoid the highly corrosive environment acid environment. Superacid catalysts prepared with $AlCl_3$ and H_2SO_4 or poly(styrene sulfonate) (PSSA) were prepared and characterized for acid hydrolysis of PET at 150 ^oC. The impact of catalyst type, composition and concentration on PET conversion and TPA yield were monitored. The recovered TPA was characterized with FTIR and NMR to compare with commercial TPA . The PSSA based acid/super-acid catalysts can be recovered for reuse thereby avoiding the acid disposal issue in usage of homogeneous acid catalysts.

Topics:
Catalysts
Environmental Sustainability

Invited Talk: Catalytic Upcycling of Polyolefins

- Conference: Global Symposium on Waste Plastic
- Year: 2019
- Proceeding: 2019 Global Symposium on Waste Plastic
- Group: Abstract Submissions
- Session:
  Session 3: Reuse and Recycle of Waste Plastic
- Time:
  Friday, September 20, 2019 - 10:45am-11:10am

Authors:
Jenks, C., Ames Laboratory

With the majority of plastics used to date being single-use and because of the lengthy time for their natural degradation, new pathways are needed to efficiently and economically re-use these materials. Chemical transformation of waste plastic to create higher value products, known as upcycling, is an emerging area of research. Low-density polyethylene recycling, for example, is extremely limited at this time, and the potential of upcycling these materials could be part a multi-pronged solution for handling existing waste and moving away from single-use plastics. Here, we discuss pathways to catalytically deconstruct waste plastics with a focus on retaining their inherent chemical and energy value. We consider the advantages and disadvantages of this approach compared to recycling and converting back to monomers or syngas. As part of a multi-institutional collaboration, Argonne National Laboratory is designing new, cooperative catalysts to *selectively*, rather than randomly, cleave polyolefins. A key advantage of this approach is that relatively uniform macromolecules can be produced avoiding the formation of light alkanes through excessive hydrogenolysis.
Topics:
Catalysis

Invited Talk: Chemical Recycling of Polyethylene Waste By Oxidative Degradation

- Conference: Global Symposium on Waste Plastic
- Year: 2019
- Proceeding: 2019 Global Symposium on Waste Plastic
- Group: Abstract Submissions
- Session:
  Session 3: Reuse and Recycle of Waste Plastic
- Time:
  Friday, September 20, 2019 - 10:20am-10:45am

Authors:
Le Roy, J., BioCellection

Today, only 9% of plastic packaging gets recycled worldwide. The rest goes to landfills, incinerators, and
oceans. At BioCellection, we aim to protect our environment through creating innovative recycling processes for currently unrecyclable post-consumer waste plastics by converting this waste into virgin quality building blocks for sustainable supply chains. Our process is developed for polyethylene, which is
over a third of all plastics produced globally. We are currently focused on LDPE and HDPE flexible plastics - think grocery bags, bubble wraps, trash bags, retail packaging, food wraps, etc.

We have developed a thermal oxidation chemical recycling process that enables polymers to be broken
into lower molecular weight species with oxygenated terminals, forming valuable organic acid compounds that can be harvested, purified, and used to make products. Compounds created from our process include succinic acid, glutaric acid, adipic acid, pimelic acid, suberic acid, and azelaic acid. These
are the first chemical intermediates made from post-consumer waste! Today's intermediates are produced using petroleum, and they're essential in the production of performance materials, solvents, coatings, and more. Our innovation unlocks the potential of using plastic waste to replace fossil fuel as a
resource for sustainable supply chains.

Topics:
Chemicals & Materials

Combining Reclaimed PET with Bio-Based Monomers Enables Plastics Upcycling

- Conference: Global Symposium on Waste Plastic
- Year: 2019
- Proceeding: 2019 Global Symposium on Waste Plastic
- Group: Abstract Submissions
- Session:
  Session 3: Reuse and Recycle of Waste Plastic
- Time:
  Friday, September 20, 2019 - 11:25am-11:40am

Authors:
Nicholson, S., National Renewable Energy Laboratory

Less than 30% of PET plastic is recycled in the United States.[1] The low reclamation rate may be due to mechanical recycling accounting for the majority of commercial PET recycling. Mechanical recycling yields a lower-quality recycled plastic than the process of chemically recycling back to the monomer phase, which can be energetically and economically expensive. In this work, recently published in *Joule*, reclaimed PET bottles are chemically recycled and converted into higher-value fiberglass-reinforced plastics (FRPs).[2] This presentation will focus on the supply-chain level energy and greenhouse gas emissions intensities associated with this novel upcycling route. The Materials Flows through Industry supply chain analysis tool, developed at NREL, is used to estimate these impacts.[3] Depending on the extent and speed of adoption, this strategy for plastic upcycling could lead to significant energy usage and emissions reductions over the traditional FRP supply chain. Overall, this approach could provide an economic incentive for plastics recycling and renewable feedstock use through the creation of higher-value FRPs.

References

[1] National Association for PET Container Resources, and the Association of Plastic Recyclers. Report on postconsumer PET container recycling activity in 2014. https://www.plasticsrecycling.org/images/pdf/resources/reports/Rate-Reports/APR_NAPCOR_2014RateReport.pdf

[2] Rorrer, Nicholas A., Scott Nicholson, Alberta Carpenter, Mary J. Biddy, Nicholas J. Grundl, and Gregg T. Beckham. "Combining Reclaimed PET with Bio-based Monomers Enables Plastics Upcycling." *Joule* (2019).

[3] Hanes, Rebecca J., and Alberta Carpenter. "Evaluating opportunities to improve material and

energy impacts in commodity supply chains." *Environment Systems and Decisions* 37, no. 1 (2017): 6-12.

Abstract:
P581492.pdf
Topics:
Energy (Sustainability & Environment)
Industrial Biotechnology

Plenary Talk: The Basel Convention, Plastic Waste and the Circular Plastic Economy

- Conference: Global Symposium on Waste Plastic
- Year: 2019
- Proceeding: 2019 Global Symposium on Waste Plastic
- Group: Abstract Submissions
- Session:
  Plenary Presentations

Authors:
Ternes, M. E., Earth & Water Law, LLC

On May 10, 2019, at the 14th meeting of the Basel Convention Conference of the Parties (COP), participating nations agreed to add "plastic waste" to the list of "other wastes," effective January 1, 2021. The addition of "plastic waste" to the Basel Convention means that the global trade in plastic waste will be regulated subject to the Convention's protections requiring legitimate recycling, and a new Partnership in Plastic Waste will begin work on solutions. And even though the United States has never ratified the Basel Convention, the United States is directly impacted, as countries are now turning away U.S. shipments of plastic waste and sending it back.

With this global circular plastic economy, the flow of new plastic into commerce will be affected, impacting manufacturers and consumers alike. The global effort to adapt to a cradle-to-cradle plastic economy necessarily contemplates collective engagement by government, manufacturers, consumers and recyclers alike in order to fundamentally reform the global commerce of plastic, from historically contemplated abandonment to continuing use.

Implications reach beyond the consumer plastic industry. For example, as plastic demand declines, ethylene demand may also decline, impacting hydrocarbon markets and ultimately oil and gas production and investment. Also, EPA and state regulations governing solid waste recycling may provide assistance or challenges, depending upon governmental response to recycling approaches.

The 2019 Global Symposium on Waste Plastic will provide an excellent opportunity explore implications and solutions.

Topics:
Regulations
Sustainable Engineering

Special Guest Talk: Maya Meena�s Last Straw

- Conference: Global Symposium on Waste Plastic
- Year: 2019
- Proceeding: 2019 Global Symposium on Waste Plastic
- Group: Abstract Submissions
- Session:
  Plenary Presentations

Authors:
Meena Zachry, M., Maya Meena Zachry, Dixie Magnet Elementary School

Plastic pollution is a well discussed concern as a major environmental crisis. The disastrous effects of it are already seen far and wide, drawing attention to the gravity of the situation. This alarm has directed academics, scientists, researchers and innovators alike to dedicate research and development helping manage and mitigate the environmental impacts of plastic pollution. Common perception is that such responsibility largely falls on the aforementioned parties to solve our predicament of the indestructible plastic. Another common view is that protecting and healing the earth is overwhelming and unachievable therefore an impossible undertaking by a 'non green' person. Maya Meena's Last Straw was founded in 2018 by a 3year old who passionately believes that we do not need millions doing things perfectly, we just need millions doing SOMETHING. Maya Meena's Last Straw proposes the simplicity of beginning with the plastic straw. In this presentation, Maya Meena aims to stage her oath to our Plastic Warriors committed to finding solutions, mitigating processes or innovation to manage our plastic waste that she will aid them in the means of making better choices every day so that everyday people like her and families like hers can take steps to "turn off the faucet as tightly as we can" so the true work of healing our earth can take place. Highlights of steps she has personally taken in reducing plastic waste as well as the barriers keeping her from them will be shared.

A Study of the Physical Properties of Plastic Derived Fuel Oil Produced from Waste Plastic

- Conference: Global Symposium on Waste Plastic
- Year: 2019
- Proceeding: 2019 Global Symposium on Waste Plastic
- Group: Poster Submissions
- Session: Poster Abstracts

Authors:
Browning, S., University of Kentucky College of Engineering
Kizza, R., Makerere University
Joshi, C., University of Kentucky College of Engineering
Seay, J. R., University of Kentucky
Quigley, B., University of Kentucky

Plastic waste around the globe has become a massively detrimental issue that can lead to health complications and environmental destruction. Studies and prior experimentation suggest that thermally decomposing plastic will cause the carbon bonds to break apart and form a fuel oil similar in composition to diesel. Using a low-cost and simple processor called the Trash-to-Tank (3T) processor developed by the University of Kentucky Appropriate Technology and Sustainability (UKATS) Research Team, waste plastic can be converted into fuel, known as Plastic Derived Fuel Oil (PDFO) and sold as a diesel alternative.

The 3T processor has been specifically designed to be appropriate for underdeveloped regions where access to sophisticated literature are often lacking. Currently, the UKATS Research Team is conducting a pilot study in Uganda to assess the performance of the 3T process. In order to ensure that the fuel meet all appropriate regulatory requirements for use as a motor vehicle fuel in Uganda, emissions testing has been commissioned from the Ugandan Bureau of Standards. The laboratory results for the PDFO will be presented and compared with traditional number 2 fuel oil. The 3T process has been implemented in Uganda and has the potential to provide a low cost way for rural communities in developing regions to take ownership of the waste plastic problem in their communities. This process is part of the establishment of locally managed decentralized circular economies in underdeveloped regions and further can give women and other people living in poorer regions the ability to become entrepreneurs.

Topics:
Energy (Sustainability & Environment)

Characterization of Microplastics in Precipitation

- Conference: Global Symposium on Waste Plastic
- Year: 2019
- Proceeding: 2019 Global Symposium on Waste Plastic
- Group: Poster Submissions
- Session:
  Poster Abstracts

Authors:
Anderson, B., Montana State University

The modern lifestyle is consumed by plastic. From disposable bottles to packaging to modern appliances, plastic is nearly impossible to avoid. A rapidly growing area of research is focused on microplastics, which are small plastic particles either directly produced to be small or are derived from the weathering of larger plastic pieces. Extensive research has been conducted on the characterization of marine microplastics, but relatively less is known about freshwater systems and especially transport throughout the water cycle. Due to the ubiquitous nature of plastic, there are likely microplastics in precipitation and they can be found in remote environments. This research project aims to characterize microplastics presence in Montana precipitation. To complete this task, precipitation samples (i.e. rain, snow, hail) were taken, filtered, and observed using epifluorescence microscopy to count and categorize plastic particles. Raman spectroscopy was employed to chemically identify unknown polymers. Thus far, preliminary results have shown plastic particles in precipitation. The findings of this research project will further inform the prevalence of microplastics and their role in environmental contamination. As society moves further towards sustainability and environmental responsibility, it is imperative that we actively work to understand and resolve the implications of our plastics use.

Topics:
Environmental Sustainability

Circularity, Cost and Life Cycle Impact: Accounting for Trade-Offs in Choosing Plastic Grocery Bags and Waste Management Strategies

- Conference: Global Symposium on Waste Plastic
- Year: 2019
- Proceeding: 2019 Global Symposium on Waste Plastic
- Group: Poster Submissions
- Session: Poster Abstracts

Authors:
Thakker, V., The Ohio State University
Bakshi, B. R., The Ohio State University

With the onus of waste mitigation and elimination of oceanic plastics, it is essential to establish Circular Economy for Plastics. This work aims to develop a theoretical framework for the assessment and design of alternative plastic products while accounting for Circularity, Impact and Costs. A holistic network is developed to link various modules in life-cycles of five different types of plastic bags that can be used by a household. A Multi-objective non-linear optimization problem is solved to find the globally optimal pathway in this network. Optimization criteria include - Degree of Circularity, Global Warming Potential and Cost. A novel metric captures the restored value in down-scaling of waste plastic and loss of natural capital due to littering, thereby describing the economic circularity of the network. Decision variables are the fraction of each bag type used, preferred collection method and waste management strategies.

Two stake holders are considered for the objectives, namely - Individual Households and Society, corresponding to Bag costs and Life-Cycle Costs respectively. Pareto fronts for three objectives are plotted and an optimal set of decisions is proposed, which, from Societal perspective is the uptake of reusable low-density polyethylene (LDPE) bags and single-use high-density polyethylene (HDPE) bags, with curbside collection and combination of recycling and incineration. Whereas, the decisions from perspective of consumers shift towards using a fraction of bags which are biodegradable (poly-lactic acid - PLA), and partial waste management by composting. Numerous additional insights are obtained from the study, including sensitivity of results towards re-usability of bags.

Topics:
Computing and Systems Engineering
Product Design
Sustainable Engineering

Development and Analysis of a Waste Plastic Generation Profile for Kampala, Uganda

- Conference: Global Symposium on Waste Plastic
- Year: 2019
- Proceeding: 2019 Global Symposium on Waste Plastic
- Group: Poster Submissions
- Session: Poster Abstracts

Authors:
Turkmani, R., University of Kentucky
Joshi, C., University of Kentucky College of Engineering
Seay, J. R., University of Kentucky

Pollution is a significant global challenge. Municipal solid waste accumulates on land in many areas globally due to lack of appropriate disposal methods, capital and infrastructure for managing waste. In some rural areas, waste management solutions are nonexistent, meaning waste piles around residential locations. One sustainable solution to this problem considers complete elimination of plastic from the ecosystem. This solution is termed as Trash to Tank (3T), a process which converts waste plastic types (polystyrene, polypropylene, low-density polyethylene, and high-density polyethylene) into Plastic Derived Fuel Oil (PDFO) via pyrolysis. The fuel produced is similar in composition to diesel and kerosene.

A country suitable for use of 3T is Uganda, lacking sufficient capital and infrastructure to manage all of its waste. In order to optimize the utilization of 3T process, reduce waste plastic accumulation on land and in the natural waterways of Uganda (River Nile and Lake Victoria), a waste plastic generation profile must be developed and analyzed. This study accomplishes this for a small subset of Uganda, the capital city of Kampala. Through partnership with Kampala Capital City Authority (KCCA), a local waste management municipality, existent waste plastic generation was determined for the city. This information was correlated with available data on population density and geographical analysis of income level via ArcGIS (a spatial analysis tool) to understand the impact of population and income on waste generation. The results of this study reveal the interplay of these factors for waste plastic management in Kampala and similar sub-Saharan African cities.

Topics:
Environmental Sustainability
Fuels
Sustainability & Environment

Emissions Analysis of Plastic Derived Fuel Oil Using Trash to Tank Process

GLOBAL SYMPOSIUM ON WASTE PLASTIC

- Conference: Global Symposium on Waste Plastic
- Year: 2019
- Proceeding: 2019 Global Symposium on Waste Plastic
- Group: Poster Submissions
- Session:
  Poster Abstracts

Authors:
Quigley, B., University of Kentucky College of Engineering Paducah Campus
Joshi, C., University of Kentucky College of Engineering
Seay, J. R., University of Kentucky

Plastic is ubiquitous in our environment. Along with it comes significant waste plastic generation, which is an imminent threat to both terrestrial and marine ecosystems. A potential solution for completely eliminating waste plastic from the environment is a process called Trash to Tank (3T), which converts waste plastic to a fuel similar in composition to diesel and kerosene via thermal decomposition in the absence of oxygen. This fuel is termed as Plastic Derived Fuel Oil (PDFO) in this research study.

In this analysis, an appropriate technology based 3T process, which utilizes an Instove Rocket Stove and fabricated stainless steel retort were used to perform slow-pyrolysis (temperature of 250-400°C) of plastic types, high-density polyethylene, low-density polyethylene, polystyrene and polypropylene to PDFO. However, for PDFO to be widely used as a diesel or kerosene substitute, or as a blend for petrodiesel, its emissions must be tested. Therefore, utilizing a diesel engine and a Bacharach combustion analyzer, the nitrogen oxide, sulfur oxide and carbon monoxide emissions of the various PDFOs were analyzed. The results of this study are presented in this research project.

Topics:
Sustainability & Environment

Solar Thermal Depolymerization Process

- Conference: Global Symposium on Waste Plastic
- Year: 2019
- Proceeding: 2019 Global Symposium on Waste Plastic
- Group: Poster Submissions
- Session:
  Poster Abstracts

Authors:
Miao, Y., Baylor University

Plastic pollution has become one of the most pressing environmental issues, as rapidly increasing the production of disposable plastic products overwhelms the worldâ€™s ability to deal with them. Conventional thermal depolymerization process still depends on the non-renewable fossil fuels. Therefore, a depolymerization process driven by a renewable energy source (such as solar thermal energy) will become the best solution.

The proposed work will focus on the development of solar thermal depolymerization reaction set-up. The novel concept is proposed to enhance or maintain the overall efficiency of the depolymerization process without using any of fossil energy resources. The primary objective of the work proposed here is to demonstrate the feasibility of solar thermal depolymerization process with a conversion of 60~70%. The set-up includes heliostat, two-stage depolymerization reactors, heat recuperator, and distillation column. The feedstock polymers are firstly ground into small chunks and mixed with a homogeneous catalyst (Y-zeolite or $MgCO_3$) and water. It is then fed into the first stage depolymerization reactor, and the temperature and pressure are controlled at 375°C and 225bar respectively to create supercritical water. The products of this process include slurry with polymer and catalyst, which is sent to a second-stage reactor which is located inside the heliostat. The solar energy generates heat and the reaction temperature keeps at 460~500°C, breaking down the longer hydrocarbon chains. The product then flows through a heat recuperator and recycle the heat for the first stage reactor. Finally, the shorter hydrocarbons are sorted through distillation, generating oil and ethylene.

Topics:
Reactors
Solar Energy

Understanding Slow Pyrolysis of Plastic

- Conference: Global Symposium on Waste Plastic
- Year: 2019
- Proceeding: 2019 Global Symposium on Waste Plastic
- Group: Poster Submissions
- Session:
  Poster Abstracts

Authors:
Li, W., University of Kentucky
Zhang, Y., University of Kentucky
Shi, J., University of Kentucky

Plastic pollution is a growing environmental concern worldwide. Recently, there are growing interests to convert plastic into varieties of advance materials with tailorable properties through slow-pyrolysis. However, unlike fast pyrolysis of plastic which has been extensively investigated, the process of plastic slow pyrolysis is so far not well understood. Therefore, in the present study, slow pyrolysis of plastic was investigated with a commercial pyrolyzer – GC/MS system. The overall product distributions, including gas, volatile and solid products were tracked at different heating rates (2, 20, 40°C/min) and different temperature regions (100-200°C, 200-300°C and 300-600°C). Solid residues were further characterized for potential material application to determine the morphology, pore structure, conductivity, interfacial chemistry, and mechanical properties using SEM/in-situ TEM, XRD, BET, FT-IR/Raman, XPS and solid phase NMR.

Topics:
Energy (Sustainability & Environment)
Nanomaterials

Upcycling of Polyethylene Terephthalate Using Petase and Deep Eutectic Solvents

- Conference: Global Symposium on Waste Plastic
- Year: 2019
- Proceeding: 2019 Global Symposium on Waste Plastic
- Group: Poster Submissions
- Session:
  Poster Abstracts

Authors:
Hunter, J., University of Kentucky

Plastics are becoming a major waste product across the world. Plastic upcycling involves depolymerizing plastic into its monomers for reuse and it has the potential to create a new avenue in plastic recycling and new market for value-added products.

Polyethylene terephthalate or PET is a thermoplastic polyester that is relatively chemically inert and resistant to microbial degradation. Currently, the most common method of depolymerizing PET is with expensive, environmentally harmful chemicals. However, there has been a recent discovery of a microbe that has evolved in a high PET environment and is able to degrade and metabolize PET through the use of its PETase enzyme. Deep eutectic solvent (DES) is a class of ionic solvent that incorporates one or more compounds to form eutectic mixture with a melting point lower than either of the original compounds often resulting in a liquid at room temperature. DES has low volatility and potentially better thermostability and biocompatibility, and has been shown to be able to solubilize PET. Using DES as a solvent in a PETase and PET reaction has the potential to become a cheaper alternative to the current PET upcycling methods. This study explored the biocompatibility of different DESs and PETase as well as which compatible DES has the highest rate of solubility of PET. Results from this work lead to a better understanding of the interactions between PETase and DES and help to the development of a biocatalytic plastic upcycling method using DES.

Topics:
Bioprocessing

Using Life Cycle Thinking to Understand the End-of-Life Impacts of Sugarcane Bagasse and Bioplastics

- Conference: Global Symposium on Waste Plastic
- Year: 2019
- Proceeding: 2019 Global Symposium on Waste Plastic
- Group: Poster Submissions
- Session:
  Poster Abstracts

Authors:
Newby, D., University of Kentucky

In 2018, the Government of Belize introduced an implementation Strategy and Action Plan to phase-out plastics as well as Styrofoam and to transition to green products like bioplastics. Bioplastics are plastics derived from plants, one of which is sugarcane bagasse (a byproduct from sugar production) plastics. Given that Belize agriculture exports consists of 60% sugarcane, bagasse is produced in excess. In this study, we will investigate the environmental impacts of replacing petroleum single-use plastics with sugarcane bagasse plastics as well as using anaerobic digestion to manage bioplastic waste. We propose to use life cycle assessment (LCA) to assist Belize with strategies for eliminating single-use plastics and methods for managing the bioplastic waste via anaerobic digestion. By using LCA, we will evaluate the environmental impact of using the bioplastics as an alternative in addition to the impacts of utilizing anaerobic digestion as a waste mitigation technique. We will consider the end-of-life cycle processes of the bagasse such as the collection, transportation, and disposal in addition to the offsets such as energy, fertilizer and biochar to analyze its environmental impact. Through analyzing these processes, we will be able to evaluate the environmental impacts of replacing single-use plastics with bagasse plastics and management of the bagasse plastic after use. Anaerobic digestion of bioplastics could lead to many different applications where the energy produced could make sugar production self-sustaining, biochar could be used for buffer zones to stop river pollution and fertilizer could enhance soil quality.

Topics:
Biofuels (Energy)
Sustainability & Environment
Sustainable Engineering

AUTHOR INDEX